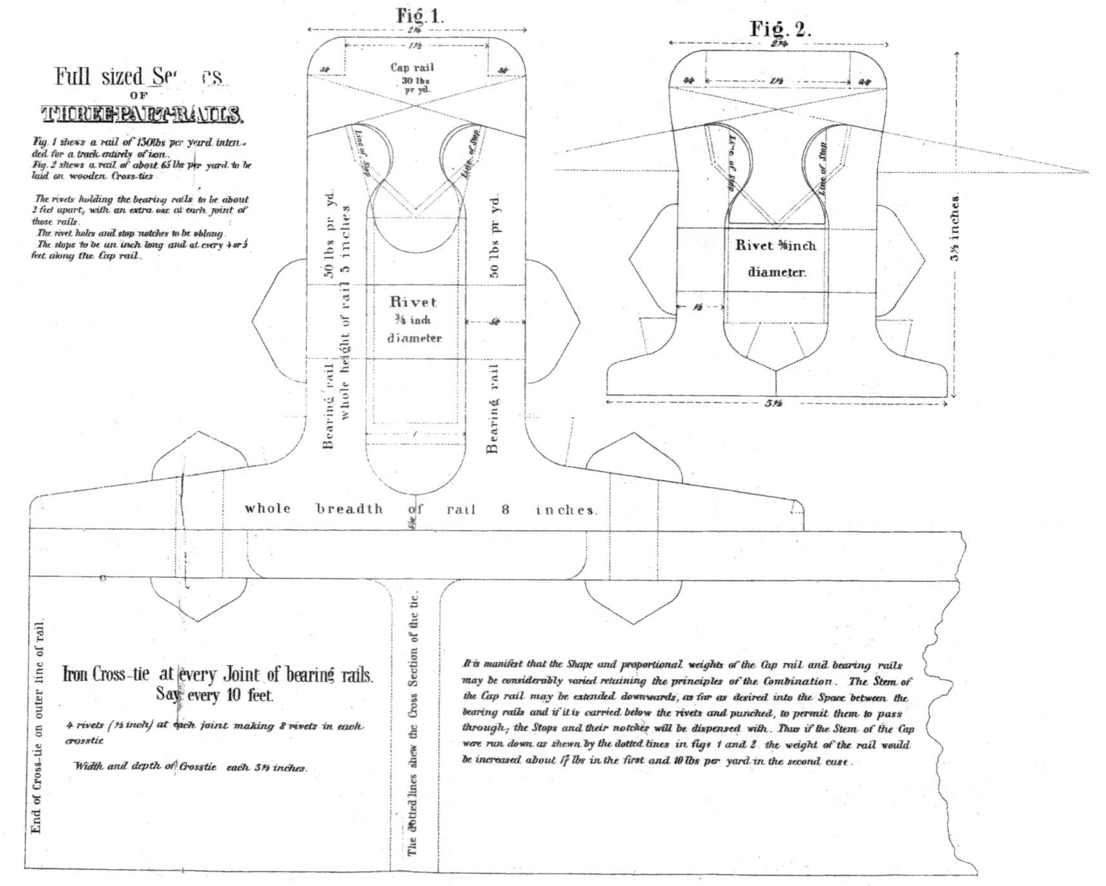
Full sized Se ... es
OF
THREE PART RAILS.
Fig. 1 shews a rail of 150lbs per yard intended for a track entirely of iron.
Fig. 2 shews a rail of about 65 lbs per yard to be laid on wooden Cross-ties.
The rivets holding the bearing rails to be about 2 feet apart, with an extra one at each joint of those rails.
The rivet holes and stop notches to be oblong.
The stops to be an inch long and at every 4 or 5 feet along the Cap rail.
Fig. 1.
Cap rail
30 lbs pr yd.
Line of Stop
50 lbs pr yd.
Bearing rail
whole height of rail 5 inches
Rivet
¾ inch
diameter
whole breadth of rail 8 inches.
Fig. 2.
Rivet ⅝ inch
diameter.
3½ inches.
Iron Cross-tie at every Joint of bearing rails.
Say every 10 feet.
4 rivets (½ inch) at each joint making 8 rivets in each crosstie
Width and depth of Crosstie each 3½ inches.
End of Cross-tie on outer line of rail.
The dotted lines shew the Cross Section of the tie.
It is manifest that the Shape and proportional weights of the Cap rail and bearing rails may be considerably varied retaining the principles of the Combination. The Stem of the Cap rail may be extended downwards, as far as desired into the Space between the bearing rails and if it is carried below the rivets and punched, to permit them to pass through; the Stops and their notches will be dispensed with. Thus if the Stem of the Cap were run down as shewn by the dotted lines in figs 1 and 2 the weight of the rail would be increased about 17 lbs in the first and 10 lbs per yard in the second case.

REMARKS

UPON THE DEFECTS OF RAILWAY TRACKS

AND THEIR REMEDY.

Although the railway structure, in its simple elements, is not an invention of modern times, (the Egyptians are supposed to have used it,) and although in its more matured form, it is now upwards of twenty years old, yet it is still in a progressive state, and is admitted on all hands not to have attained perfection, but to be marked by some serious defects. The best evidence of this, is the great variety of opinions which still prevail in regard to the details of its form and combinations. There is but little agreement among professional men, even in the leading principles of the structure—that is, in regard to the section of the rail, the mode of supporting it, the manner of connecting it at the joint, &c. The undersigned has been an attentive observer of the constant agitation to which these questions have been subjected, and has, as he believes, carefully and impartially weighed the arguments for and against the various ways proposed for the accomplishment of the object which all have had in view, viz. a firm yet somewhat elastic structure—well connected at the joints of the bars and other points of contact of the different pieces composing the track, and yet readily taken apart in the process of repair. His conclusion has been, that there were radical defects in all the existing systems of form and combination of parts, defects to be remedied only by a fundamental change in some of the principles of the structure, and, in his view, the cure is as clear as the cause of these imperfections.

A railway track has always heretofore consisted of a series of iron bars, of greater or less length and of various sectional forms, laid together at the ends, and supported throughout their lengths in various ways. Care has usually been taken to give these bars a sufficiency of strength in themselves, and their supports, to make them form a solid and smooth surface for the rolling of the wheels upon them throughout their length—and if the bars could be incorporated together by welding at their ends, or in other words, if there were no joints in the track, there would be little left to desire in any of the well built railways of the present day; for, although the resistance of the soil on which the track rests not being uniform, undulations will occur in the surface of the rails—yet there would be no abrupt depressions or elevations—and the carriages would oscillate with an easy swinging movement, attended by no concussion. The rails, however, not being continuous, but terminating at every distance of 15 or 20 feet, the wheel has to pass from one to the other, over a very narrow gap it is true, but one quite wide enough almost to annihilate the resistance of the bar at that point as a *beam,* and to make it depend for its power to hold up the wheel, principally upon the resistance of the substance supporting it at that spot, whether that substance be a longitudinal bearer or cross sleeper of timber, or a stone block—according as one or other of these three systems of support are used. Now to give the bearer, or sleeper, or block, the resistance necessary to compensate fully the loss of strength as a beam, which the bars sustain at their ends, has been in practice found *impracticable*—although an approximation thereto may be had by increasing the compactness of the road-bed and the extent of surface bearing thereon at the joints. The approximation, however, is always an uncertain, and, at best, an imperfect one, and even in the case of the continuous timber bearer under the rail, which would appear to give the best support to the joints of the bars, the result is unsatisfactory; the compressible character of the wood always permitting the rails to sink into it more or less at their ends. The proved impossibility of effectually sustaining the joints by increased compactness of road-bed and provision of additional bearing surface, has induced attempts to connect the iron bars at their points of contact so as to form a *splice,* for the restoration of the strength lost at this point. But all such efforts have proved abortive, the effect of the quick passage of heavy trains being to shake and wear loose all the fastenings constituting the splice, and it is now quite apparent that little more can be effected at the joints than to keep the rails, vertically and laterally, sufficiently in place to permit the wheel to pass safely from one bar to another—and not always safely indeed, for there are not wanting instances wherein disastrous accidents have been occasioned by the escape of the rail from what confined it at this point. The only real splice that has been thus far applied to the joints of a line of rails, is the continuous bearing timber; but this, in consequence of the yielding of the wood, is much impaired in its effect, as already observed. I make the preceding statements as well known facts, not denied by any one at all experienced in the construction and maintenance of railways, or an observant traveller upon them. I could give, were it necessary, a mass of detail upon the subject, collected in my own visits to various lines of railway, and in my correspondence with my professional brethren all over the United States. Suffice it to say, that the existence of the evil of bad joints, and the difficulty of dealing with it, is amply demonstrated by the variety of expedients to *palliate* what is conceded to be *incurable.* Upon one line, for example, are to be found two cast iron chairs weighing together 45 lbs., applied to each bar—and upon another line, no chairs at all, but the rails simply let into the sleepers at the ends and spiked down. Upon other roads a chair weighing from 15 to 25 lbs., with a wooden key to hold fast the rail in it. Upon others again, chairs weighing from 12 to 20 lbs. with a lip on each side to lap tightly over the bottom flanges of the rail. On others a simple plate to support the ends of the bars and keep them sidewise in place, depending upon spikes to hold them down. On others the same plate with screw bolts in place of spikes, and lastly, upon one or two roads, two splice plates fitting on each side into the hollows of the rail and drawn tight by screw or cotter bolts in a horizontal position. None of these modes of making the joints secure, operate as a splice, in effect, or but partially, at the first; for the violent blow given to the end of the bar by every passing wheel, soon jars loose the firmest grip that these fastenings can take. The wooden keys, besides their swelling and shrinking as the atmosphere changes, soon loosen, and the continual driving of them up rapidly wears them out. The spikes break, the screw bolts either snap off or have their thread stripped—and the keys of the cotter bolts become bent and broken from driving them up. The tight fitting clamp chair either breaks or wears loose in its lips. In short, no fastening that can be applied at this point holds its own long—and where, in despair of getting any contrivance to stand, all are rejected, and the rail is simply nailed down upon the sleeper, the latter is soon deeply indented and rapidly worn away by the bars it supports. These mischiefs are of course most observable in the lines of heavy traffic, and some time in use. Upon the great trunk line between Boston and Albany, opened in 1842, the clattering and thumping of the joints was deafening when I passed over it in 1848; the chairs and ends of bars having become loose in their fit, from the wear of six or seven years, and on the same line large renewals of rails had then, and are now, annually taking place, chiefly in consequence of the effect upon the ends of the bars of the blows received there.

In describing the previous modes of fastening joints the form of rail is supposed to be one of the only two sections now used in America, the T or H and the ∩ or bridge rail, each with a broad base, supporting itself. The plain T rail, or the double headed rail, (I,) requiring a chair to support it at every bearing, have been long since discarded here; the latter indeed was never used in the United States, and I believe less in England now than formerly—the bridge rail seeming to be, at present, the favorite pattern there. All that is said of the joints of the H and ∩ rails, is, however, fully applicable to the other sections, the tracks laid some twelve years ago, in the United States, with the T rail, held in a chair by keys, became, before they were superseded by another form of rail, the roughest and most dangerous railways ever travelled over. A passage over them, was, indeed, terrifying at last, although when first laid down they were pretty smooth, (vide the Columbia, the New Jersey, and the Eastern rail roads before their reconstruction.) Not having had the good fortune, as yet, to be wafted over an English railway, the undersigned has no personal experience in regard to them—but he knows that they are subject to the same evil which he has described as effecting our American roads—and although the great attention paid in England to the adjustments of the track, together with the superior ease and comfort of most of the English first class carriages, and the fact that passengers are confined to the inside of the cars, (with the windows closed,) during their journey, and not allowed, as here, to stand at times upon the outer platforms; although these circumstances have prevented travellers from noticing the shock and noise in passing the joints, it certainly must and does exist, as is testified by observant persons who have travelled upon the railways of England.

It should be mentioned, while speaking of the difficulties of maintaining the joints of a track, that the endwise movement of the rail, under the blow of the wheel, is one of the most prominent. The bars are not all operated on alike by this cause, and consequently, some being pushed farther than others, the openings at the joints become irregular, some being closed entirely, and others widely open. This movement of the rail is a very dangerous one, and unless watched, would soon force the rails quite out of their chairs, especially upon double lines of railway, where the movement of the trains, on each track, is always in the same direction.

The evils have been now, perhaps, sufficiently descanted upon, and we will proceed to speak of their remedy.

Instead of making the rails in solid or single bars, laid end to end, it is proposed to make them in parts combining to form the cross section of the rail—these parts breaking joint with each other, and held together by rivets, so as to form, in effect, a continuous bar of compound structure—being as near an approach to an unbroken line of iron as it is physically possible to make. The necessity of providing for contraction and expansion, and for repairs to those parts of the rail, which, from the unavoidable inequalities in the texture of the metal, will require to be renewed at different times, manifestly renders the welding of the bars together, at their ends, impracticable. We must then break up the body of the iron into parts, which may shrink and dilate, and be removed and replaced, independently of each other; and the question is, how this division of the mass shall be effected? Hitherto it has been by simply cutting off the line at intervals—the mischiefs of which mode have been fully shown. The other mode, now proposed, is to divide the mass not only cross-wise, but longitudinally—that is, not only in length but in section, making the several parts, resulting from this sectional division, unite at different points in the length of the line—so that the wheel will always be sustained by the full strength of one portion of the rail, while passing over the gap occasioned by the division of another portion. It is, in short, nothing more than the application of the "break joint" principle to rails, so long recognized, and so successfully used in carpentry and framing of every kind, and indeed, in the railway itself, by those who prefer the continuous bearing timber, to the cross sleeper or the stone block. By this division of the bar, a portion of its extreme strength is relinquished, the compound bar being, at its strongest point, somewhat weaker than the solid bar, in the middle of its length.—But as the maximum of strength is less in the compound bar, so the minimum is much greater, and thus that approach to uniformity of strength is effected, which is the desideratum; a structure presenting an alternation of very strong and very weak points, being of all others the worst. It has been said, that the compound bar is somewhat weaker in its strongest point, (which would be the point midway between any two contiguous joints,) than the single bar in the middle of its length.—But the difference would be less considerable than might be supposed. I judge from an experiment made with a compound bar, of 50 lbs. per yard, and a solid bar of similar weight, but of the bridge pattern. The two bars with a clear bearing of three feet, and under a strain of three tons, applied to the middle of their lengths, exhibited the same deflection, although the compound bar had one of its joints, (in one of the under sections,) between the bearings. I do not quote the experiment as conclusive. It was but a single one, and may have been under the influence of some circumstance, not observed, which gave too favorable a result for the compound bar. But while conceding fully, that there is a reduction of the maximum of strength, the more than corresponding increase in the minimum, already claimed for the composite rail, is an invaluable acquisition. The loss of extreme strength, is, however, not the only objection that might naturally be urged against the compound rail, and, if that compound principle has been already thought of in England, its practical application has probably been prevented by the objection now to be mentioned. I refer to the supposed difficulty of connecting, in a substantial and satisfactory manner, the parts into which the bar is divided, so as to make them hold well together without shaking, or working, or breaking, or looseness of any kind, and so, also, as to permit them, at the same time, to yield freely to the effect of changes of temperature. In reflecting upon the subject, I am so strongly impressed with the superiority of the compound principle, as to feel *amazed* that it has not, long ago, been adopted, and can only account for the fact that it has not, by this last consideration, which, I confess, operated awhile upon my own mind, so as to make me in the outset, not entirely confident of success in introducing the composite rail. My professional friends, indeed, nearly all hung back at first upon this ground—admitting that the idea in the abstract was a happy one; but fearing its defeat by the supposed impossibility of holding the parts properly together. Nothing short of experimental demonstration would be satisfactory in this, more than in other cases. The fact that the fundamental principles, involved in the structure of the rail, were altogether in favor of what was aimed at, did not seem to me to make a due impression. Because the fastenings, by which it was attempted to connect the solid bars at their ends, failed to perform their intended functions, so must the attachments of the compound bar! But, the distinction in the two cases, consists simply in the prevalence in the two plans of two opposite principles; in the solid rail the principle of *concentration,* and in the compound rail the principle of *diffusion*. In the former, the whole shock experienced in passing the point of separation of the bars, is concentrated at one point, the single joint—in the latter, the one great shock is divided into two or three very much lesser shocks, and thus is diffused with a greatly diminished intensity over the whole length of the bar, and the fastenings being also similarly dispersed along the bar, the scattered and softened concussions are effectually resisted by them. It would be difficult, indeed impracticable, to compare the different momenta of the mass which experiences these shocks, in the cases of the two kinds of bars, as to do so, with precision, it would be necessary to know exactly the relative spaces through which the wheel of the carriage would fall, in passing the single joint of the solid bar, and the two or three joints of the compound bar. There can, however, be no doubt that the sum of the momenta, in the latter case, would fall much within the single momentum in the former case, and that, therefore, the carriage and the rail would sustain much less injury from the shocks occurring in passing over the compound bar.

But, leaving the mathematics of the matter, I will go to the more practical views to be taken of the subject, and to the light which the experience, which has happily been obtained, casts thereon. I have thus far discussed the principle of the compound rail in the general, and without reference to any of the various forms it may assume. But I will now say, that notwithstanding the tardiness with which most of the Engineers of this country, to whom the improvement has been submitted, have seemed disposed to yield their assent to its value, there is a growing feeling in its favor, sufficiently evinced, by the fact, that there have been already no less than four different forms of compound rail, proposed by various parties connected with the making or management of railways. Of these several forms, three are double rails, composed of two parts, with splicing pieces at the joints, and the fourth, suggested by myself, consists of three parts, without other splicing. Now, if we admit that the composite rail is, in the abstract, the best form, we have then to choose between the possible varieties of this form, that may be suggested. The reflection given by the undersigned to this selection, has led him to a preference of *three* parts, to any other number of parts, for the combination—although, if there were no other alternative, he would take the rail of two parts in preference to the single rail. It is sufficiently evident that there would be no temptation to increase the number of parts to more than three.

It is not proposed to discuss, in detail, the relative merits of the three forms of two part rail above indicated. Their suggestions sufficiently shows the tendency of professional opinion, in this country, on the general subject. Were I compelled to choose between the three rails, I would take the simplest and most symmetrical of them. But over any form of rail composed of two parts only, I consider the one of three parts possesses the following advantages: 1st. The rail, at its weakest point, will possess nearly two-thirds of the strength of the solid bar, while the two part rail cannot, (independently of the splice, which could as well be applied to the rail of three parts,) have more than half its full strength at that point. 2nd. The pressure of the wheel is communicated to each half of the base of the rail more equally through the cap or third part of the rail. 3d. This cap, by means of the dove-tailed rib upon its under side, adjusts the bearing rails so as to bring their tops always to the same plane, and any inequality in their height is seen at their bottoms only, where it is of the least importance. 4th. The cap rail so connects the bearing rails by hanging them, as it were, upon each other, by means of the entrance of the lips of the latter, into the side grooves in the bottom rib of the former, as to apply no cross strain to the connecting rivet, which is not the case with the two part rails, unless by a tongue and groove in the halves of the two part rail running their whole length, the bolt may be also relieved of cross strains, or a filling-in piece, occupying the hollow of the rail, would produce the same relief, provided it were possible to make the tongue and groove or the filling in piece, fit every where with precision. This, however, is impracticable, and hence, the advantage of the dove-tail rib, which, when the bearing rails are drawn together by the rivet, adjusts the bearing of all the parts in contact, upon each other. 5th. The cap rail being separable from the bearing rails, may be removed and replaced independently of them, and as the entire wear takes place upon the cap, which would never constitute more than from $\frac{1}{3}$ to $\frac{1}{4}$ of the entire rail, the remaining $\frac{2}{3}$ or $\frac{3}{4}$ contained in the bearing rails, may be regarded as enduring in perpetuity. This property may, indeed, be realized to some extent, in a rail composed of two parts; but not to an equal degree—for in any two part rail, the upper or cap rail would, of necessity, have to bear a much larger proportion to the whole rail, in order to give it body sufficient to ensure a safe connection with the bottom part. In the three part rail, the cap may be made as light as is consistent with its wearing well. I do not, however, after much reflection upon the possible forms of a two part rail, divided *horizontally*, perceive how a satisfactory and permanent connection between the upper and lower halves can be effected, as the dependence must be upon the resistance of the rivets, (or bolts or keys if substituted for rivets,) to a strain—which they cannot resist long, but will soon wear loose if they do not break.

It is obvious, that if there be a sufficient advantage in filling the hollow of the rail with wood or iron, it can be done in the rail of three parts as well as, or better, than in that of two—but I doubt the utility of such an addition to the combination, which would be rendered more expensive and complex thereby, without, probably, a corresponding improvement in the strength or stability of the structure. If wood, however, be introduced, it should be seasoned so as to shrink as little as possible, and if protected against decay, by the kyán or other analogous process, it would be all the better. It should then be shaped to the section of the cavity, and made very slightly larger than that, so that when the rivets were inserted, they would, in shrinking, draw the bearing rails together, and the cap rail downwards, so tightly as to compress the wood both horizontally and vertically, and thus produce a very close fit in all the parts of the combination. It will be apparent that a filling-in piece of iron or a tongue and groove, could as well be applied to the three part as to the two part rail—but I prefer depending on the self-adjusting action of the cap rail for the connection of the parts of the system. I think I have now discussed the subject upon theoretical principles at sufficient, and perhaps unnecessary, length, and will give a brief statement of such facts relating to it, as are in my possession.

It is upwards of four years since, after several previous years' reflection upon the feasibility of constructing a track entirely of iron, without wood or stone supports, the form of compound rail of three parts, exhibited in the drawings hereto attached, suggested itself to me. I had, more than three years ago, an experimental piece of track, 200 feet in length, constructed of cast iron, and laid in the yard of the station at Mount Clare, in the environs of Baltimore. The cap rail, as well as the bearing rails, were of cast iron. The weight of cap was 17⅓ and of each bearing rail 42—total of the three 101⅓ lbs. per yard. So were the cross ties of cast iron, and they weighed 37 lbs. each, and were 10 feet apart. The whole was held together by screw bolts, ½ inch diameter, and two feet apart. The track was laid upon sand ballast, without other support of any kind. It has been ever since daily passed over by the heaviest trains that come at slow speed into the yard, and has received scarcely any attention in the way of adjustment. Yet it has held together perfectly, and although the cap rails, which were very light, and, like the bearing rails, cast in lengths of 20 *feet long*, have crumbled down at the edges, as cast iron always will under the tread of railway carriage wheels, but one or two of the bearing rails have broken and but one cross tie. The success of this track encouraged me to look to the general application of the principle in *rolled iron*, to rails of any weight over 50 lbs. per yard, and with or without supports of timber—as the size and weight of rail might make necessary. As I had commenced with what I then considered the maximum of weight, (108 lbs. per yard,) in the rail, I made the next trial with the assumed minimum, (of 50 lbs. per yard,) and having an opportunity of getting the rail rolled in the neighborhood of Baltimore, I had enough made to lay a section of 600 feet, and afterwards other sections of 200, 900, and 4500 feet, upon different points of the line. It was laid upon cross sleepers alone, placed about 2½ feet from centre to centre, and 7½ feet long, and 5×6 inches in section. The ballast, broken stone and gravel. A part of the first section was fastened by bolts, ½ inch in diameter, the remainder was rivetted. The bolts held very well, but required some occasional attention to make sure that they were tight. The rivets needed no such attention, and were, consequently, preferable. The cap rails were prevented from moving endwise by small key plates put through the neck of the rib, and with their ends fitting into notches, cut in the tops of the bearing rails. For the manner in which these experimental sections of track have worn, and how they have carried the trade and travel, for different times, during the last year and a half, I refer to the accompanying copies of certificates, signed by various officers of the Baltimore and Ohio Rail Road Company. I need add nothing more than to say, that every additional month which goes by, only adds to the favor in which the new track is held by those who have charge of it, or travel over it. I am, therefore, enabled to appeal to that authority from which there is no further appeal—*experience*, and that acquired under circumstances so little favorable to the new track, that the argument is entirely *a fortiori*, when we reason about the results of future trials under more suitable conditions, for success.

In figure 1, of drawing appended, I have assumed a rail of maximum weight, with a view to its being laid directly on the ballast, without wood or other support. I consider this entirely admissable, on general principles, and as having been also tested by the experience of the iron track laid at the Mount Clare Station, as already mentioned. The bearing surface of each line of rails upon the ballast is 8 inches in width, which is the same with that of the wood and iron track of the Baltimore and Ohio rail road, the substructure of which consists of a longitudinal timber under each rail but 8 inches wide—while the iron track I propose, has vastly more strength and stiffness, and thus diffuses the pressure on the ballast much more lengthwise than the other. This consideration of longitudinal diffusion of pressure is not generally attended too sufficiently. If you have two tracks of equal bearing surface on the ballast, but one of which is twice as well connected and stiff as the other, you have virtually double the bearing surface on the former. If we compare the bearing surfaces of different tracks, we find that they vary a good deal. In America, in tracks of 4 feet 8½ inch guage, laid on cross sleepers only—there are generally 6 sleepers of 7½ feet long, by about 7 inches wide, allowed to every 15 feet. This gives 1¾ square feet of bearing surface per linear foot of track. The tracks laid with longitudinal timbers are few in number here; but their bearing surface per linear foot is from 1⅓ to 2 square feet. In England larger provision is made for bearing surface, and it seems generally to range from 2½ to 3 square feet per linear foot. It may be questioned how far we should go in increasing the bearing surface, as the more extensive it is the more difficult and imperfect is the operation of packing under it. In this view I should prefer a very stiff rail with small bearing, to a flexible rail with a large bearing—and if I am right in supposing that the rail of the model I propose is nearly or quite twice as stiff as any of the strongest rails in use, (at least at and in the neighborhood of their joints) then my rail would be in effect, as well supported by the ballast as the others. It should in this connection also be noticed that the bearing surface of a track is effective in proportion to its *proximity* to the line of the rail—and thus the bearing obtained by longitudinal timbers is really worth more a good deal than that of cross sleepers. Now the English railways which have most bearing surface are laid on cross sleepers. The London and North-western, has 3 square feet per foot run. The Great Southern and Western of Ireland has 3⅜—the first is a 4 feet 8½ inch guage, the second 5 feet 3. The Great Western of England has 2½ square feet per foot, and is of 7 feet guage. The Midland Great Western of Ireland (of 5 feet 3 guage) has 2 $\frac{7}{12}$ square feet per foot. The two first are laid only on cross sleepers—the third upon longitudinals, and the fourth upon a combination of the two. These particulars have been obtained from the July (1849) Number of the Civil Engineer and Architects' Journal, wherein is an article on the subject of

"Permanent Way"—of which I shall presently make further use, and which seems to have been prepared and published very opportunely for the present purpose. In this article, the author, Mr. Dockray, proposes an improved track for the London and North-western road, as the result of the investigations recently made by him, under the instructions of his Company, into the subject of "Permanent Way" with a view to the removal of the admitted defects existing in their own and other roads. His report is a very interesting paper and is confirmatory of all the views I have expressed in relation to the faults of the present system of railway construction. The improved plan which he proposes is a good one; perhaps as good a combination of wood and iron as could be suggested: but it is a combination of two materials so different in their properties that they cannot be made to act in harmony with each other. All the changes which they undergo are different in their nature and degrees. While hot weather shrinks the wood it swells the iron—and while damp weather swells the wood, it does not affect the iron. The wood is soft while the iron is hard, and flexible while the iron is stiff—consequently the weak points of the iron cannot be strengthened by the strong points of the wood, so as to affect a perfect compensation. The effect of these diversities in the materials will always inevitably be to disarrange and ultimately destroy the combination. On the other hand, the iron track is perfect within itself as a *system* and depends for a support only upon the ballast on which it rests, and which constitutes no element in the combination. And even if wood is interposed between the iron and the ballast, it is in the way of a *rest* only and not as a part of the system of the track.

I will now present an estimate of the cost of the iron track I propose, and will compare it with those of the several tracks treated of by Mr. Dockray, and which he says he considers "the best of their kinds." I will take, in this estimate, the scale of prices employed by Mr. D., although they are higher than those at present prevailing.

Estimate of Cost of Constructing 15 *Feet in Length, of Single Line, laid with "Three-Part Rail"—exclusive of Labor in Laying Down the Track.*

No. Parts.	*Description.*	*Weight.*	*Rate.*	*Amount £*	*s.*	*d.*
6	Pieces, making 2 Rails, 130 lbs. per yd.	1,300 lbs.	at £10 per ton, ...	5	16	1
2	" Cross Ties, 75 lbs. each,	150	do. do.	0	13	4
18	Rivets, ¾ inch diameter, ½ lb. each,	9	at 4*d.* per lb.	0	3	0
16	" ⅛ do. do. 1-6 do.	2⅔	at do.	0	0	11
	Punching Rivet Holes, cutting Stop Shoulders and Notches, and straightening and fitting Rails for laying, at 1*s.* per yd.			0	5	0
42	Parts, and Estimated Cost of 5 yards, ...			£6	18	4
	Do. do. 1 do.			£1	7	8
14.784	Parts, &c. do. do. 1 mile, ...			£2434	13	4
	Or about $11,900 per mile—(U. S. currency.)					

If we compare this estimate with those of the six different plans presented by Mr. Dockray, we will see that it is but about £70 per mile more costly than the Great Western, of which it has but about half the number of parts. That it is £126 per mile cheaper than the improved plan of Mr. D., and has but few more parts—and finally that it is but £167 per mile more than the average cost of the 6 tracks including those the least improved. If the cost of laying down the road was included in all the estimates, I apprehend that my plan would compare more advantageously, as it would certainly be more easily laid, the riveting being a rapid and cheap operation and the bedding of the rails and cross ties on the ballast being much more readily done than that of the cross sleepers and longitudinals, including the dressing and adjusting of them.

I have estimated the cost of punching rivet holes, &c., from the results of actual experience in those operations in the 3 part rail I have laid, allowing in full for the superior size, and weight of the present rail. I have also supposed each of the parts of the rail to be but 15 feet long, and two cross ties to each 15 feet length—whereas in practice I would make the rails 20 feet long, and thus save one-fourth of the joints and cross ties, and a proportion of the rivets also, as well as in the number of *parts* per mile sufficiently to reduce the latter to the same with that of the simplest form of track, (Sir John Macneill's method,) viz. 36 parts per 5 yards, or 12,672 parts in a mile of single line.

But lest it should be doubted whether the extent of bearing surface of my track would be sufficient, and whether timber could be altogether dispensed with, I will suppose, that instead of the cross tie of iron, every 7½ or 10 feet, there is a cross sleeper of wood every 5 feet, or 3 for every 5 yards—these sleepers being 9 feet long and 6×10 inches—containing 3¾ cubic feet—costing 5*s.* 6*d.* each—and amounting to 16*s.* 6*d.* This item then would stand in place of the 13*s.* 4*d.* of the iron ties—the spikes, which would be used in lieu of the cross tie rivets, about balancing the rivets, and the cost of punching the holes for them. Thus, by the substitution of wooden for iron cross ties, the cost of the track per 5 yards would be increased 3*s.* 2*d.*, or per mile £55 15*s.* and the bearing surface would be increased from 1.5 to 2.83 square feet per linear foot of track. The advantage of the rivets accompanying the iron cross tie would be foregone; but this might, perhaps, be considered as compensated by the additional bearing surface. On the other hand, a perishable material would be introduced into the structure, and the excellent characteristic of a track entirely of iron would be given up.

It is manifest that the bearing may be increased at pleasure by multiplying the cross ties, but in a track so strong and stiff there could be no motive, I think, for this increased expense of construction, except where extraordinary bearing was demanded by very soft sub-soil. I think, also, that the iron cross tie yields a support and connection to the joints of the bearing rails which is highly valuable, and I should be averse to giving it up—and if more bearing surface were regarded as indispensable, I would obtain it by inserting wooden cross sleepers at points intermediate to the iron ties. The *cant* of the rail with the iron tie is made by curving the tie up at the ends sufficiently for the purpose.

I have no means of comparing the cost of maintaining a "permanent way"—upon my plan with that of any of the various existing modes of construction in England. I am very confident, however, that the difference in favor of the former would be very great indeed. From observations thus far upon the track laid with the 50 lb. rail, of which mention has been made, I am satisfied that not less than a third of the labor of adjustment will be saved, and the renewal of materials should be in at least as favorable a proportion. An inspection of the rail will show the facility with which any one of the 3 parts composing it may be removed and replaced—all that is required for this purpose being the cutting off the rivet heads with a chisel. But this will be an operation rarely required, and the cheapness of the rivets makes the cost of material a matter of little consequence. If I am right in my suppositions, then the superior safety and smoothness of the new track, attended by a considerable reduction in the cost of repairs to engines and carriages, and a great increase of public security and comfort and consequently an accession to the popularity of railways as a means of travel, would all combine to place the value of the improvement in a very conspicuous position.

It will be perceived that in my estimate of the cost of the *rails* of the new track, I assume the same rate as for a solid bar. This might be objected to, as new patterns are always made an excuse by the manufacturers for asking higher prices, and might lead, in the beginning, to some slight difference of cost, but I have the authority of the manufacturers who made the 50 lb. rail for me, for saying, that the cost of manufacture will be no greater than that of the bridge or T rail.

The bearing rails (weighing about 50 lbs. per yard each,) will be very readily manufactured. They are in fact nothing more than "angle iron." The cap rail (of 30 lbs. per yard) contains a feature in it which would appear to make it difficult to roll, but I think this can be managed very readily; and if not, why the feature is not at all indispensable, and, indeed, but slightly, if at all, important to the plan. I refer to the slight pitch outwards and downwards of the under sides of the cap where it rests on the bearing rails on each side of the central rib. It is evident that the rail could be rolled with these slopes in it, only by nearly completing it without them in the first place, and making them in passing it through the final groove of the rolls, leaving enough metal on the sides of the cap to press down into and fill up the triangles between the slopes and the horizontal line above them. The quantity of metal requisite for this is so small that I think it can be done easily; or it might, perhaps, be as well or better done in one or two other ways I need not describe. But if it cannot be done at all, it matters but little. The only advantage of the pitch is, that it relieves the rivets a little—but the rivets will hold perfectly well without this relief, as is known now from the experience of the 50 lb. rail, in which not a single rivet has broken under a very heavy traffic, for five months past, although the rivets are but $\frac{5}{8}$ inch diameter, and the rail so light, while the ballast not being well packed in the first place, and the sub-soil soft, there has been a good deal of irregular settlement and consequent strain on the parts. In the cap of the rail I have put into the neck of the rib, on each side, what I call a "stop." This is to keep the cap rail from moving endwise—the stop fitting into a notch cut in the lips of the bearing rails. I propose that this stop shall be made in the process of rolling, and it can be effected by cutting a notch in the tongue of the rolls into which the metal will press up, by a reverse action to that of the *nipple* which makes a countersink in a rolled bar. The stop as it comes from the rolls would not be square at its ends, for the notch in the tongue would have to be wider at top than bottom to let the metal press in and pass out easily. The shoulders of the stop must then be cut square afterwards, with a chisel, and so will be cut the notches in the lips of the bearing rails. The stops may be as close as desired—they cannot be farther apart than the circumference of the rolls, say 4 or 5 feet, and this, I think, would be about the right distance for them. But if they could not be conveniently made in the rolling, they could be put in afterwards by punching a square hole through the neck of the rib and putting in a plug of iron, hot, like a rivet, having metal enough to make the stops on each side of the rib, and dressing it into shape with the chisel. Or instead of the stops, the key plates already mentioned, as being used with the 50 lb. rail now laid, may be employed. They will hold the caps very well, but the stops would be preferred, and especially if they can be made in rolling the cap rails. All the rivet holes and the stop notches are made oblong, to allow for contraction and expansion. This will amply provide for it, as *experience* has shown, in the longest section of the 3 part 50 lb. rail, (4,500 feet,) which has now, for ten months, passed through all the changes of our fluctuating climate, in which the extremes of heat and cold are as great as in any other. This point which seemed one of the most to be feared, is now therefore *settled* satisfactorily, as no inconvenience has been felt from this cause.

Among the merits of the compound rail, will be apparent that of retaining its line in curves, better than the solid rail; the breaking of the joints producing in the bearing rails a mutual counteraction of the tendency to straighten into chords, after being sprung to the curve. With bars so long as 20 feet, it is supposed that, even in the heaviest patterns, it will not be necessary to *set* them, by previous bending, which, in the solid rail, would be indispensable. The additional strain upon the rivets of the compound rail, will not be objectionable, as it will be in the direction of their length only, and much within their power of resistance.

A very remarkable advantage from the division of the rail into parts will also be the improvement in the *quality* of the metal. The disposition to increase of weight, has been checked by the difficulty of making a heavy bar perfect. It is understood that the rails of 100 lbs. per yard recently rolled have turned out so indifferently as to induce a return to lighter patterns. However this may be, it is quite certain that a single bar of any weight cannot be made as sound and tough, as two or three bars of the same length and aggregate section. The compound principle will permit the tendency to increased weight of rail to go much farther than would be possible in the single rail.

If it should be apprehended that the detached cap of the three part rail will not wear as long as the top of the solid rail, it is answered that this is not necessarily so with a well-proportioned cap-rail—and if the separation of the upper portion of the section from the lower should tend to this result, it should be counteracted by the better texture of the lighter bar. Experience, however, thus far indicates no greater wear in the cap rail than in the upper surface of solid rails in use for the same time.

To those, however, who are best disposed to admit the truth of the above remarks, it will be of interest to know what the new rail will cost in the first instance, compared with other common forms of track, and, for information on this question, the following estimates of cost for a rail of medium weight, to be laid in the most usual way, upon cross-ties of wood, are offered to show how the three part rail will, in general, compare with the solid rail in expense of construction :

Solid Rail—65 lbs. per yard.

1. *Rails*—20 feet long—$102\frac{2}{3}$ tons, at $60 per ton,	$6,158	
2. Joint fastenings of any variety of form, at 75c. per joint, for 528 joints, ...	396	
3. 2112 cross-ties, $7\frac{1}{2}$ feet long—6×6—laid $2\frac{1}{2}$ feet apart, at 20c., ...	422	
4. 9504 spikes, 3 to the lb—3168 lbs. at 5c. . .	158	
5. Laying track, materials, and ballast being delivered, viz. spreading ballast, bedding and dressing cross-ties, laying and fitting rails and joint fastenings, spiking, adjusting, and trimming track, at 65c. per rod of $16\frac{1}{2}$ feet,	208	
Total estimated cost of one mile of track, exclusive of ballast, ...		$7,342

Three part Rail—65 lbs. per yard.

1. Rails 20 feet long, $102\frac{2}{3}$ tons, at $60 per ton,		$6,158	
2. Rivets, 6336, $\frac{5}{8}$ inch diameter, at 4 to the lb., 1584 lbs., at 6c.		95	
3. Keys, 1056, $2\frac{1}{4}\times1\frac{1}{2}\times\frac{3}{8}$ inches, at 3 to the lb., 352 lbs., at 6c., ...		21	
4. Punching holes for rivets and key, and fitting rails for laying, at,	65c. per rod,		
5. Riveting rails after laying, . .	16 "		
6. Leveling and dressing cross-ties,	17 "		
7. Spreading ballast, . . .	17 "		
8. Spiking rails, ...	5 "		
9. Fitting, adjusting and trimming track	10 "		
Total of the above items per rod of track, ...	$1 30 per mile,	416	
10. 2112 cross-ties $7\frac{1}{2}$ feet long, 6×6, laid $2\frac{1}{2}$ feet apart, at 20c.		422	
11. 9504 spikes, (one extra at every joint) 4 to the lb. 2376 lbs., at 5c., ...		119	
Total estimated cost of one mile of track, exclusive of ballast, ...			$7,231
Difference in favor of the three part rail, per mile, ...			111

The 4, 5, 6, 7, 8 and 9th items of the last estimate are derived from the actual experience of laying the three part rail already put down, under circumstances not at all favorable to economy. It is believed that at least 10 per cent. on the aggregate of these items could be saved in future work of the same kind upon a large scale. The entire cost of laying the solid rail track is taken at the sum of the 5th, 6th, 7th, 8th and 9th items, which is certainly favorable enough to that track. This work having generally cost at least 50 per cent. more. The cost of joint fastenings for that track is assumed at 75 cents, with less than which, a tolerably good and safe joint cannot be made, although many tracks have been laid with much cheaper joints.

The items of both estimates making up the cost of *workmanship* show the *nett* expense—to which a fair profit for the contractor should be added in preparing estimates for actual construction. The prices of all the other items include cost of delivery and profit upon the articles.

The detail in which the estimates are given will enable any one to apply them to particular cases. For a lighter or heavier rail, the cost of fastenings and workmanship would differ little from those of the rail of 65 lbs. weight here assumed.

It will thus be sufficiently manifest that the three part rail will cost *no more* than any other rail of the same weight. With this admission, its friends may be satisfied, for its other advantages must prove its superiority, and, ultimately, it is believed, ensure its adoption.

The accompanying sketch, in figure 1, shows the rail of 130 lbs. per yard proposed for a track entirely of iron. Fig. 2 represents a rail of 65 lbs. per yard, to be laid upon timber supports. Other weights of rail from 50 lbs. upwards can readily be proportioned so as to carry the principle into effect with an advantage increasing with the weight, and it may be said of the compound rail, especially in this form, that as its division into parts, gives it, at all times, an elasticity which a solid bar of the same weight cannot possess, so it will retain that elasticity with a weight which would make the solid bar too rigid, except at the joints, where all such bars are alike weak, and the heaviest the most so in comparison with their strength in the middle of their length.

The subject has now been sufficiently discussed upon its general merits, and the facts and arguments above presented are offered to the profession, soliciting consideration and not shunning criticism. The author is but one of the laborers in this important field of improvement, and has argued the merits of the compound principle in general terms, and so far in favor of all the forms it may assume; and although decided in his preference of the three part pattern, he will be glad to see the suggestions of others, subjected to the test of experiment.

BENJ. H. LATROBE, Civil Engineer.

APPENDIX.

The undersigned, masters of road and supervisors of repairs upon the parts of the Baltimore and Ohio Rail Road, on which the several sections of the new rail composed of three parts have been laid—unite (each for himself in regard to the part of the track under his own charge) in the following statement of their experience and opinions of the merits of that rail.

There are four different points at which the rail has been laid down, viz. 200 feet in the Mount Clare cut, near Baltimore, laid in May, 1849, and now down ten months—600 feet, a mile west of Avalon Works, laid in May, 1848, and now down twenty-two months—900 feet, at Sykesville, laid in May, 1849, and 4,000 feet, near Cumberland, laid in June, 1849. All of these sections have been laid upon cross-ties only, without any sub-sill, excepting the piece near Sykesville, where a sub-sill was used, the embankment being newly made. The weight of the rail is about 50 lbs. per yard. It is riveted together and spiked down upon the cross-ties without chairs, joint plates, or other fastening.

We are enabled to say that the track thus far has given us great satisfaction, and that we are led to consider it a decided improvement upon any form of railway structure of which we have any knowledge. The manner in which the pieces composing the rail break joint with each other, and the simple and permanent mode in which they are connected, equalize the strength of the track, so that it forms, in effect, a continuous rail without joints—while there is an entire freedom from the shock and clatter which invariably takes place at the joining of the bars of other tracks, there is, at the same time, a general elastic spring in the new rail, which relieves the passing train of any harshness, or jar, in the movement, and which must diminish, as we suppose, the wear and tear of engines and cars, as much as it does that of the track.

We find that considerably less labor is required to keep the new track in adjustment than the old, under similar circumstances, and we should feel much more safe in leaving it in an imperfect state of adjustment than the other—the joints of which require constant attention.

We see no reason thus far to think that the top or cap rail will wear more rapidly than the upper surface of any solid rail—the few bars in the section near Avalon, which have shown some wear at the ends, were defective when laid down (the cap being also of the lighter pattern, afterwards made heavier for use in the other sections) nor is there in that section more signs of wear than are to be observed on other similar lengths of the ∩ or T rail in use for the same time.

The rivets appear to hold the rail together well, and of those of the larger size, few or none break when the track is well laid and settled. The expense of renewing rivets must, in any event, we think, be very trifling compared with that of replacing the chairs, plates, and bolts, spikes, or other joint fastenings of other forms of rail. The spikes of the new track will also, we believe, outlast those of the old rails; and spikes of a smaller size will answer the same purpose.

We do not believe there will be any difficulty in promptly rebuilding the track if deranged by the running off of trains, and we think that, connected as it is, from end to end, it will be very difficult, if, indeed, possible, that a train should tear it to pieces as it does a track composed of solid bars, depending only upon the ordinary fastenings at the joints to hold them together. The three part rail now laid is of the lightest weight suitable to a rail of any form laid on cross-ties; and we judge that as it has, notwithstanding, succeeded so well, it would show its good properties still more strongly if it were of any heavier weight. There appears to be no difficulty in the accommodation of the rail to the contraction and expansion of the several parts, by heat and cold. Although the track may be still regarded as an experimental one, only because it is comparatively new, yet we think it has been long enough in use to remove all the objections we have heard or felt against it, and we are satisfied it embraces principles of construction which distinguish it from all preceding tracks, and which must ensure its ultimate success, as a highly valuable improvement upon the railway structure.

WENDEL BOLLMAN, Mast. of R. 2d Div. B. & O. R. R.
WILLIAM D. BURTON, Supervisor.
ROSEBY CARR, Supervisor.
JAMES B. JORDAN, Foreman shops, Mt. Clare Depot.
S. T. SHIPLEY, Master of R., 1st Division, B. & O. R. R.
ROBT. MURRAY, Supervisor.
THATCHER PERKINS, Mast. of Mach'y, B. & O. R. R.

March, 1850.

The undersigned, employed in various capacities upon the passenger and tonnage trains of the Baltimore and Ohio Rail Road Company, being requested to express their opinions upon the merits of the new form of rail composed of three parts, and of which several short sections have been laid upon the Company's road during the past year, unite in the following statement:

The movement of the engine and cars of the train upon this rail is more smooth and free from shocks and jar than upon any other form of track with which they have any acquaintance upon this or any other road. There is also much less noise either from the track itself, or from the train. The solidity of the track is at the same time accompanied by an elasticity and softness of motion which they suppose must diminish greatly the wear and tear of the machinery, and the permanent connection of all the parts of the rail, appears entirely to obviate the risks of accidents from the displacement of the bars. In short, the undersigned best express their views in regard to the track, by saying that they feel *safer and more comfortable* upon it, than upon any other, over which they have passed.

JAMES CLARK, Engineman,
ELIJAH COLLINS, Engineman,
HENRY FLURY, Carpenter,
WILLIAM OWENS, Conductor Passenger Train.
ALBERT WATTS, Engineman,
DARIUS DARBY, Conductor,
H. G. DAVIS, Conductor Passenger Train,
J. B. FORD, Agent at Cumberland,
REUBEN WATTS, Engineman,
THOMAS BECKETT, Engineman,
JOSEPH BROWN, Supervisor of Trains B. & O. R. R.
ISRAEL C. YORK, Master Engineman,
WILLIAM BUTLER, Engineman,
WILLIAM D. WINTERS, Fireman,
GRAFTON DARBY, Engineman,
G. G. FRETHY, Conductor,
THOMAS PURDY, Fireman,
D. KOONCE, Conductor,
GEO. W. RICHARDSON, Engineman,
CHARLES WALTEMYER, Fireman,
JOSEPH BRUCHEY, Engineman,
OTHO REEL, "
ADAM RIDDLE, "
MARINUS CHANCE, Fireman,
JOHN C. JACOBS, Engineman,
BENJAMIN F. SILVESTER, Engineman,
ORMOND BUTLER, Engineman,
J. H. HANN, Conductor Passenger Train,
JOHN SCOTTI, Engineman,
WILLIAM W. POLLOCK, Engineman,
PERRY MITCHELL, Blacksmith.

www.ingramcontent.com/pod-product-compliance
Lightning Source LLC
LaVergne TN
LVHW020636110826
845149LV00004B/1215